New England Forests Through Time

7/04

This Book was purchased
with funds from a
Department of Conservation
and Recreation
Grant

New England Forests Through Time

Insights from the Harvard Forest Dioramas

DAVID R. FOSTER
DIRECTOR, HARVARD FOREST

JOHN F. O'KEEFE
COORDINATOR, FISHER MUSEUM

PHOTOGRAPHS BY JOHN GREEN

HARVARD FOREST / PETERSHAM, MASSACHUSETTS
HARVARD UNIVERSITY

DISTRIBUTED BY HARVARD UNIVERSITY PRESS
CAMBRIDGE, MASSACHUSETTS, AND LONDON, ENGLAND

Library of Congress Cataloging-in-Publication Data
Foster, David R.
    New England forests through time : insights from the Harvard Forest
Dioramas / David R. Foster, John F. O'Keefe ; photographs by John Green.
        p. cm.
    Includes bibliographical references (p.).
    ISBN 0-674-00344-6 (pbk. : alk. paper)
    1. Forests and forestry — New England — History. 2. Forest
conservation — New England — History. 3. Forest management — New
England — History. 4. Harvard Forest (Research facility) 5. Diorama —
Massachusetts — Petersham. I. O'Keefe, John F. II. Title.

SD144.A12 F67 2000
634.9'0974 — dc21   99-058912

PRINTED IN CHINA

This book is dedicated to
Richard T. Fisher and Ernest G. Stillman,
who inspired and oversaw the creation of the
Harvard Forest dioramas,
and to
Samuel Guernsey and Theodore Pitman and their colleagues,
whose artistry fulfilled the vision.

CONTENTS

FOREWORD

In the late 1920s, Richard T. Fisher — a Harvard professor of forestry and the director of the Harvard Forest — and a philanthropist, Dr. Ernest G. Stillman, both dedicated conservationists, conceived a plan for an ambitious exhibit depicting the natural history, land-use history, and ecology of New England forests. Their vision was for a series of dioramas — miniaturized but realistic three-dimensional scenes — showing different aspects of forest history, ecology, and management, to be housed in a museum at the Harvard Forest in Petersham, Massachusetts. Visitors to the museum could learn directly from the dioramas and then take these lessons outside to the woods and demonstration trails that lead through the Forest's 3,000 acres.

Through the dioramas Fisher and Stillman sought to capture the essence of the evolving Harvard Forest approach to environmental science, in which a sound understanding of landscape history provides a basis for interpretation and conservation of nature. This approach has provided the grounding for many students since 1907, when the Forest was established by Harvard University, and it continues to provide the inspiration for new research directions in forest ecology at the Forest. This historical-ecological approach has also proved applicable to modern environmental issues as it becomes increasingly apparent that changes in nature can only be assessed through long-term perspectives.

To achieve their vision, Fisher and Stillman enlisted the Cambridge-based studio of Guernsey and Pitman, a firm with great experience producing small-scale and highly detailed models for museums and libraries at Harvard University. The Harvard Forest project challenged these artisans to work at an unusually large scale compared to their previous work, and to achieve an unprecedented level of biological realism in comparison to other museum dioramas of the day. For example, the artisans had to create not only realistic-looking miniature trees of various species, each species with its unique form, branching pattern, and bark characteristics, but also entire woodland scenes with the appropriate groupings of vegetation, wildlife, and landscape features such as laneways, stonewalls, and farmsteads.

The result of this three-way collaboration among a scientist, a philanthropist, and the artists was a series of dioramas unlike any others created before or since. In their lifelike detail and unsurpassed ecological veracity, the dioramas show the history of the New England landscape over the past 300 years, lessons from the conservation history of the region, and approaches to the management of the resulting forests. Collectively they provide fascinating visual lessons drawn from the region's natural and cultural history by Harvard Forest scientists.

New England Forests Through Time

Landscape History of Central New England

Seven of the Harvard Forest dioramas form a historical series that depicts changes in the New England landscape over the past 300 years at one location. The scene was designed by Richard Fisher to depict all the important transformations of the landscape in the upland area of central Massachusetts since the pre-settlement period.

Shortly after beginning their first studies in 1906, Fisher and his students recognized that an understanding of the history of human activity and natural disturbance was crucial to the interpretation of the ecology of the New England landscape and to the management of the region's forests for conservation, timber, or wildlife value. Consequently, Fisher initiated a series of studies of the remaining old-growth forests in central New England as well as comparative investigations of the post-settlement history of the region and its forests.

The first of the historical dioramas is a scene depicting the pre-settlement forest, based on studies at the old-growth Pisgah forest that Harvard University owns in southwestern New Hampshire. Subsequent scenes, based on Harvard Forest's studies of post-settlement events, show the progression of the same site through intensive agricultural use to farm abandonment to subsequent stages of natural forest succession and active forest management. This pattern of human activity and forest change has characterized most of the New England countryside in the few centuries since European settlement. The seven historical dioramas capture the major stages of this fascinating cultural and ecological story and thus help us to understand and interpret the present New England landscape. This legacy of 300 years of change — a major factor shaping many details of the modern land — is apparent to those who walk the forests today and learn what clues to look for.

The ecological and historical interpretation of the details of this remarkable transformation of the New England landscape, and the assessment of their significance, have changed little since Fisher and his colleagues designed the dioramas in the 1930s. Importantly, this early research activity laid the groundwork for many of the current approaches to ecology, conservation biology, and forest management at the Harvard Forest and beyond. Consequently, one of the major strengths of the research and educational program at the Forest today is this legacy of nearly 100 years of intensive historical and modern studies, backed by a comprehensive archive of long-term data.

PRE-SETTLEMENT FOREST
1700 A.D.

The pre-settlement landscape was not a stable, homogeneous, unchanging forest. Rather, it showed considerable temporal and spatial variation in the mixture and distribution of species and the pattern of vegetation. An ongoing process of natural disturbance — by hurricanes, other windstorms, ice storms, pathogens, and fires ignited by lightning strikes — led to differences in the age, density, size, and species of trees across a wide range of sites, from protected valleys to exposed ridges. Variation in soils and water availability were other factors controlling both the pattern and dynamics of the landscape: sites flooded by beavers and areas inundated by annual fluctuations in the water table supported herbaceous or shrub vegetation in contrast to the surrounding forests. Humans also altered the pre-settlement forest: Indians made clearings for their villages and fields; across the landscape they burned extensive tracts of forest to improve hunting. The resulting mosaic of diverse habitats supported a wide range of plant and animal species and offered the European explorers and settlers abundant new opportunities as well as many difficult challenges.

The diorama depicts some of this variation in vegetation, ranging from rather open forest on rock and shallow soils to a diverse old-growth forest of hemlock, beech, oak, birch, and pine. The presence of paper birches on the left indicates that the forest is dynamic, as this species requires abundant light, such as occurs in an opening created by fire or an extensive blowdown of trees, in order to establish and grow.

At the time of European settlement the forest landscape of central New England was largely covered with mixtures of broad-leaved and coniferous trees. North-central Massachusetts occupies what is called the Transition Forest Zone; here, the northern hardwood–hemlock–white pine forest found in much of Vermont, New Hampshire, and southern Maine overlaps with the central hardwood or oak-hickory forest, which extends southward to the mid-Atlantic states. The cool moist conditions found in the northern part of the Transition Zone, in cool ravines, on north slopes, and at higher elevations favored the growth of "northern" species, including hemlock, beech, yellow birch, sugar maple, poplar, red spruce, and balsam fir. More exposed and drier sites on ridges, well-drained soils, and to the south supported "southern" species, including white, black, and scarlet oak, hickory, chestnut, black birch, and pitch pine. Intermediate sites were characterized by a mixture of both groups, and in addition by white pine, red oak, white ash, black cherry, and red maple.

Forests dominated by white pine occurred primarily on dry, sandy soils and in dynamic environments such as pond shores, beaver meadows, and abandoned Indian fields and villages. In such places, full exposure to light and a temporary reduction of competition from other tree species permitted the pine's abundant reproduction. Detailed analysis of Harvard's old-growth Pisgah forest in Winchester, New Hampshire, has shown that hurricanes, other windstorms, and fires have disturbed the forest periodically through time. These infrequent disturbances created a mosaic of even-aged pine and pine-hemlock-hardwood forests and enabled shade-intolerant species such as paper birch and black cherry to persist along with the more shade-tolerant hemlock and beech. Many stands were established after some natural disturbance.

PRE-SETTLEMENT FOREST 1700 A.D.

AN EARLY SETTLER CLEARS A HOMESTEAD
1740 A.D.

For most of the New England region, European settlement occurred largely during the eighteenth century, the date being progressively later to the north and in the rugged mountainous areas. In much of southern and central New England in the first half of the eighteenth century, the Colonial government granted land to groups of individuals, called proprietors, to enable them to establish new settlements in the frontier wilderness. In what was to become the town of Petersham, Massachusetts, where most of the Harvard Forest is located, the government granted a 36-square-mile territory to 60 families, but required them to settle on it within three years, in order to retain the land and incorporate the town.

The settlers came from "frontier" towns farther east in Massachusetts such as Rutland, Lancaster, Lunenburg, and Brookfield. They left their old homes early in the spring in order to clear as much forest as possible before winter set in. At least for the first year or two, they traveled back and forth seasonally between their old homesteads to the east and their new clearings in the wilderness, where, for immediate shelter, they often built makeshift lean-tos on their rough land. But clearing land produced a lot of logs, and in Petersham — and probably in many other such towns — the first sawmill was established very soon after settlement.

Clearing involved felling trees and burning over the land. When selecting land to clear, the settler determined the quality of the soil by careful inspection of the forest composition, noting, for example, that sugar maple, beech, and ash grew on productive sites that were suitable for crops; chestnut and oak dominated the broad, moderately productive uplands that made good pastureland; and hemlock and red maple sites were usually wet, rocky, and less suitable for agriculture. First, the choicest timber trees — usually white pine, pitch pine, oak, and chestnut — were cut into logs to be sawn into boards and planks or hewn into timbers for frame construction. With a good supply of pine boards and framing material the construction of permanent dwellings began, and within three to four years an early settler could boast of a small house, cleared fields, small pastures, and an assortment of domestic animals, as pictured in the diorama.

A considerable amount of the wood remaining from forest clearance was split up for fuel or for fencing for the few cows and oxen a settler kept. This would still leave great quantities of wood for which there was no immediate use and little means of transportation; the excess was piled and burned as clearing progressed. Charcoal, potash, and forest leaf litter constituted valuable fertilizer from the forest for the early settlers' farm crops. There was little thought of conserving wood or wildlife, as these resources appeared to be inexhaustible. Consequently, through forest clearance, hunting, and trapping, the abundance of many species changed rapidly and the wilderness was transformed into a domesticated rural landscape.

The spring following clearing the first year's crops of corn, grains, beans, squash, and other vegetables could be planted among the stumps on burned-over land. Naturally occurring lowland meadows along lakes and streams were a good source of forage and hay for livestock. As agriculture expanded with the arrival of more settlers and the clearing of more and more land, artisans and merchants diversified their trade, and new towns rapidly developed into self-sufficient, thriving communities.

CLEARING OF A HOMESTEAD BY AN EARLY SETTLER 1740 A.D.

HEIGHT OF FOREST CLEARANCE AND AGRICULTURE
1830 A.D.

The peak of deforestation and agricultural activity across most of New England occurred during the period from 1830 to 1880. The diorama shows our settler's farm about 90 years after clearing got under way. The close network of stone walls in the whole region and the innumerable piles of stone thrown together within the clearings themselves bespeak the back-breaking effort that was required to bring the rocky forested land into good farming condition. This labor transformed New England in the late eighteenth and early nineteenth centuries into a region whose thriving economy was based on agriculture and widespread local and home industry. The land was dominated by human activity, and the population was spread fairly evenly across most of the landscape in townships of 500 to 2,000 people.

By 1840 the population of the town of Petersham was 1,775, the highest level in the town's history, and the citizens formed a prosperous community of farmers, tradespeople, and artisans who practiced a wide range of trades. This prosperity is evident in the much grander house now gracing the site. Farmers produced beef cattle, which could be easily transported — on the hoof — to distant markets, as well as a range of fruits, vegetables and essential grains such as oats, rye, barley, corn, and wheat. Across much of New England (except for northern Maine and mountainous areas), 60 to 80 percent of the land was cleared for pasture, tillage, orchards, and buildings. Most of this cleared land supported grazing animals; generally less than 10 percent was actively plowed for crops.

In nineteenth-century New England there was a great need for wood for buildings, fuel, house furnishings, and farm implements, not to mention industry. Consequently, the comparatively small remaining areas of woodland were subjected to frequent cuttings to remove the most desirable trees for lumber and the least desirable ones for fuel. Cutting, burning, and grazing of these remnant forest areas changed their composition greatly and encouraged the growth of sprouting species such as chestnut and oaks as well as light-loving early successional trees like birches and red maple. Wood was now too valuable to use for fencing, so the abundant stones were used instead.

The dramatic change in habitat combined with widespread hunting and trapping of wild animals reduced or eliminated many populations of native wildlife such as wolves, cougars, beavers, moose, deer, and turkeys. However, the open meadows, grasslands, shrublands, and young forests encouraged the expansion of open-land species such as bobolinks, whippoorwills, meadowlarks, skunks, foxes, and rabbits. Meanwhile, the construction of large dams for industry and transportation eliminated anadromous fish like shad, herring, and salmon from most streams and rivers where they had thrived.

By the mid-nineteenth century, with little woodland left and wood becoming a vital economic necessity in the new Industrial Revolution, dire predictions of scarcity abounded. But the pendulum of economic — and social — change swung yet again and relieved the pressure on local forests. Coal was introduced for heat and industrial uses; expanded sea and, later, rail transport brought lumber from Maine and from New York, Pennsylvania, and the Great Lakes states. Finally, the country's continued expansion westward resulted in a shift in agricultural activity to the Midwest and initiated the abandonment of many New England farms, which gave the forests a chance to regenerate. Meanwhile, the human population continued to grow but became increasingly concentrated in cities and industrial towns.

HEIGHT OF AGRICULTURE 1830 A.D.

FARM ABANDONMENT
1850 A.D.

Beginning in the mid-1800s and continuing for more than a century, farming declined on a broad scale across New England, as numerous forces combined to draw New England farm families away from their rocky hill lands. As farm mechanization increased, these small farms could no longer be worked profitably in competition with the rich, stone-free farmlands of the Midwest, whose products had been made more accessible to eastern markets by the construction of the Erie Canal and the railroads. The growth, associated with the Industrial Revolution, of eastern urban centers along waterways and railroads, the discovery of gold in California, and the Civil War all contributed to the precipitous decline of rural New England populations; although the total New England population continued to grow at a rapid pace, it was increasingly concentrated in urban and suburban areas. In Petersham this decline reached a nadir in 1930 when the town had less than 400 residents; even today the population is only slightly over half of its nineteenth-century peak.

The diorama shows what happened when the abandoned fields and pastures literally went to seed: they rapidly developed into forests of white pine. The great New England naturalist and writer Henry David Thoreau noted that old pine trees that the farmers had left for shade in the pastures, along fencerows, or in nearby woodlands produced an abundance of seed that dispersed easily, and then established well on the sod and grassland in

abandoned farmland. Though scattered seeds of hardwoods such as red maple, white ash, red oak, chestnut, and gray and paper birch germinated on these sites, their seedlings were more palatable than pine to grazing animals that browsed in the relatively ignored fields — and this selectively increased the pines' abundance. Two pine seed years, which normally occur every third or fourth year, ordinarily were sufficient to establish pine seedlings across fields in close proximity to a seed source. Therefore, the "old-field" pine stands were nearly always even-aged.

The new pine stands soon became exceedingly thick, so little or no undergrowth could establish itself during the next 20 to 30 years. Then, as the canopy rose, the density of stems declined due to competition; more open conditions, with increased light, developed; and many hardwoods — chiefly red maple, red oak, white ash, chestnut, black birch, black cherry, and sugar maple — began to appear, their seeds delivered to the site by wind, birds, and small mammals. The emerging seedlings and saplings gradually formed a dense thicket in the understory beneath the pines. On most sites, in contrast to the hardwoods, the relatively shade-intolerant white pine was largely unable to establish beneath its own forest canopy. Henry Thoreau described in great detail how the hardwood understory established and was poised to "succeed" (Thoreau's coinage) the white pine, should the latter be logged or damaged by windstorm.

FARM ABANDONMENT 1850 A.D.

"OLD-FIELD" WHITE PINE FOREST ON ABANDONED FARMLAND 1910 A.D.

As the "old-field" stands of white pine reached middle age, it became evident that they contained a valuable and rapidly growing crop of second-growth timber. Although the timber was often knotty and vastly inferior to old-growth white pine in quality and size, the light, easily worked wood could be used for many products — boxes, pails, match sticks, shoe and boot heels, toys, and woodenware — that were in great demand in this era before cardboard, plastics, and other packing materials. The straight, better-quality white pine timber made excellent framing material and finish lumber.

So much white pine became marketable during the period from 1890 to 1920 that portable sawmills appeared across central New England and many new wood-using factories were established. At yields of 25,000 to 50,000 board-feet (a board foot is 12 inches square and one inch thick) per acre and values hovering around $10 per thousand board feet, one might well envy a farmer who owned a 100-acre woodlot, worth perhaps $30,000. This was a huge sum at the time, to be reaped from merely having paid taxes on the idle land. Between 1890 and 1920 an estimated 15 billion board-feet of second-growth white pine, with a manufactured value of over $400 million, was cut in central New England. The peak of this timber boom occurred in 1910.

The diorama shows the clear-cutting of the small white pine stand that had developed in the farm fields abandoned after 1850. Scattered hardwoods, of little value at the time, were either cut at the base or left standing. Clear-cutting and hauling with horses was facilitated by felling the trees in strips and windrowing the slash into long lines that were burned or left to rot. The logs were carried on wooden sleds, called scoots, to a portable steam sawmill, where they would be sawn into boards and planks for easy transport out of the woods.

Once again, little thought was given to the future — in this case to ensuring a future forest or timber crop after the pine was cut. Despite this lack of forethought, another harvestable stand — of completely different characteristics — did ensue. This new forest resulted from the uncut trees and advanced growth of shade-tolerant hardwood trees that had gradually established beneath the white pine overstory. Removal of the pines allowed an even-aged rapidly growing stand to start from the sprouts and saplings of oak, black birch, chestnut, and red maple. They quickly outgrew any new pines, which, lacking the ability to sprout, had to establish as new, slow-growing seedlings. Thus the succession from "old-field" white pine to mixed hardwoods was facilitated across much of the landscape.

"OLD-FIELD" WHITE PINE ON ABANDONED LAND 1910 A.D.

"OLD-FIELD" WHITE PINE IS SUCCEEDED BY HARDWOODS
1915 A.D.

In the previous diorama, the view immediately following clear-cutting of "old-field" white pine is a sobering picture of apparently nearly complete forest devastation. Rows of heavy slash and charcoal occupy a third of the ground; scraggly hardwoods and dead pine poles have been left scattered around; piles of sawdust and worn-out camp and logging equipment litter the mill site.

However, five years after logging the scene is greatly changed, and once again the resiliency of New England forests becomes apparent. Many young hardwoods are growing in the open areas between the windrows of slash. These trees sprouted from the small stumps of saplings that had established in the shade of the pine and were cut out of the way by the choppers and teamsters when the pine was logged. Such sprouts grow extremely rapidly in open conditions, and may make straight, sound trees. In the central New England landscape the common hardwood sprouts included red maple, red and white oak, white ash, chestnut, black cherry, and black birch.

Seedlings of such pioneer, light-demanding species as gray and paper birch, pin cherry, poplar, and white pine also established themselves in the open sunlit environment after logging, but these conditions will persist less than a decade as the rapidly growing saplings and sprouts form a dense stand. In general, widespread clear-cutting increased the proportion of sprouting and short-lived pioneer species in the landscape. Also visible in the diorama are sprout clumps from the stumps and damaged stems of large hardwoods that grew up along with the pine.

Sprouts from such large stumps are generally of odd form and often have stump rot, so they die earlier than sprouts of smaller stems and are less desirable for timber.

The characteristics of the forest that grows after white pine logging depend partly on the abundance of white pine seed when the forest is cut. When a white pine woodlot is cut in a good seed year, the hardwoods are supplemented by white pine seedlings, and a few of the latter may succeed in the subsequent forest, especially on dry and sandy soils. The diorama, however, shows a case where the stand was cut in a poor year for pine seed, and therefore the newly developing forest is almost wholly of hardwoods.

In the early 1860s Henry Thoreau noted the progression from open field to pine forest and then the sudden change to young hardwood forest when the pine was cut. He called this process "succession," a term that foresters and ecologists still use. Undoubtedly, patterns of succession enhanced the diversity of forest types across the landscape, resulting in a mosaic that included young hardwood stands, "old-field" pine, agricultural fields, and older woodlots.

The vegetation mosaic of post-agricultural New England provided a wide range of wildlife habitats suitable for species of open, grassy, and shrubby areas as well as those requiring forest edges. As forests continued to expand over former agricultural land the New England landscape was poised for a return of its native wildlife.

14

WHITE PINE IS SUCCEEDED BY HARDWOODS 1915 A.D.

A VIGOROUSLY GROWING FOREST OF HARDWOODS
1930 A.D.

The diorama shows the mixed hardwood stand that followed the clear-cutting of the "old-field" white pine and has now taken definite form. One of the most characteristic features of this developing forest is that the largest trees are primarily multi-stemmed sprout clumps. Fast-growing species that sprout prolifically like red maple, gray birch, white ash, black birch, and black cherry are strongly represented.

Red oaks — one of the long-lived species that will eventually become dominant and play an important role in the future of this stand — are just beginning to overtop the other trees. In contrast to the maple, birches, and most other species, the oak will continue growing taller and wider over the next three or four decades and will gradually emerge from the general canopy of the forest to form an overstory above the rest of the trees. These dominant oaks will eventually suppress and almost stop the growth of their smaller competitors that thrived in the intense light after clear-cutting. As a consequence, there is a gradual shift in growth and wood production in these forests from a range of species when the stand was young to oak as the forest matures.

In this century red oak has assumed greater importance in New England forests as woodlands across the region have matured and as other long-lived species, such as chestnut and beech, have been reduced by introduced diseases. The formerly abundant and economically important American chestnut was lost to a fungal blight introduced into the United States from Asia in 1904. Throughout its range, ancient chestnut stumps still sprout, but once the resulting stems reach several inches in diameter they become reinfected and are killed by the blight. The roots, however, survive to sprout again. Although the introduced beech bark disease has not eliminated this important wildlife species, it has greatly reduced its abundance and vigor in many parts of its range.

In the diorama it is possible to begin to identify the developing forest type, recognize its characteristics, and identify its value for forest products, wildlife habitat, and recreation. The different growth rates of species in this actively growing hardwood forest will gradually lead to an even-aged stand of trees of many sizes, a forest that is increasing in height, tree age, and wood volume annually. Its appearance and ecological and economic benefits contrast greatly with those of the pine forest that preceded it: the diversity of trees provides a wide range of food, nest sites, and cover for wildlife and its deciduous nature produces a great contrast between summer and winter appearance. The range of wood types and qualities in this diverse forest offers both opportunities and challenges to the wood producer. As the forest ages it becomes increasingly natural in appearance, and yet the ubiquitous stone walls, woods roads, and cellar holes are constant reminders of the cultural history of the landscape.

Thus, in 100 years much of the New England countryside has gone from open agricultural land to largely closed woodland and has supported two completely different types of forest. Both forest types resulted from the interaction of human activity, variation in the environment, and the biological characteristics of the native tree species. Without planning and with little encouragement, a resilient and dynamic new forest has developed across the region.

AN AGGRADING FOREST OF HARDWOODS 1930 A.D.

THE MODERN FOREST LANDSCAPE

In the period since the dioramas were constructed the trends in forest development illustrated in the previous model (1930) have continued broadly across the eastern United States, leading to remarkable expanses of maturing forest across a densely populated landscape. Although the rate of farm and farmland abandonment has gradually declined, forest cutting has not kept pace with the growth of trees, and so the average age, size, and timber volume of forests across the region have continued to increase greatly. As these forests grow and mature and as dead and decaying wood accumulates on the ground, the forest landscape becomes increasingly natural in appearance and character. With time, too, the early successional species gradually decline and more shade-tolerant and long-lived species increase.

In this century, other natural and human processes besides land-use activities and forest regrowth have affected the forests of our region. On September 21, 1938, a hurricane, with winds exceeding 100 mph and accompanied by torrential rains and floods, cut a 60-mile-wide swath across Long Island up the Connecticut River Valley into northwestern Vermont. This storm blew down over 3 billion board-feet of timber and at the Harvard Forest approximately 70 percent of the standing volume of timber was windthrown. In order to recoup some of the economic loss and to reduce the likelihood of wildfire, the U.S. government responded to this massive disturbance by organizing the single largest timber salvage operation in the country's history.

Just as the storm dramatically altered the characteristics of the New England forests it also changed forever the perspectives of ecologists and foresters concerning the stability of natural ecosystems in the region. The tall, shallow-rooted white pines were more heavily damaged than the hardwoods, and so the hurricane hastened the regional process of forest succession to hardwood dominance. The storm also prompted considerable research, which has documented that severe hurricanes occur every 50 to 150 years in New England and that these and other disturbances and climate change lead to an ever-changing landscape.

Forests in New England have continued to be logged for a variety of wood products, and although this activity has not kept pace with forest growth, it strongly shapes the composition and structure of our forests. The forest has also been subjected to novel impacts such as introduced pests (the gypsy moth, Dutch elm disease, beech bark disease, and hemlock woolly adelgid); ozone, nitrogen deposition, and other forms of pollution; and encroachment by suburban and industrial development.

Nevertheless, New England's forests continue to thrive and expand. One indirect consequence of forest regrowth and maturation and the loss of agricultural lands has been a recent transformation in regional wildlife. Many open-land insect and plant species have become uncommon or rare, and wildlife that dominated the agrarian countryside of the nineteenth century such as bobolinks, meadowlarks, song sparrows, and quail have declined. Meanwhile, woodland plants have increased, and beavers, fishers, bears, bobcats, moose, turkeys, pileated woodpeckers and eagles have become common again.

Thus, a paradox confronts New England and much of the eastern United States: The woods and landscape are becoming wilder in appearance and in wildlife at a time when the human population and its global use of natural resources continue to increase. This situation raises major conservation and manage-

ment issues. Increasingly, people live in or near forests and often own small areas of woodlands, yet their connection to and understanding of the land, its forest, and its wildlife is at an historical low. How can humans and bears, beavers, and moose coexist in this landscape? How should we manage forests that are increasingly owned by more people in smaller units? Should we continue to import wood from other parts of the country and world, some of which are being devastated by poor logging practices, while enjoying the growing forest around us as a largely aesthetic and recreational resource, or should we obtain more wood from our home forests? These conservation issues and management questions emerge from the history of our land.

The contemporary New England landscape and the history of the countryside depicted in the dioramas afford many insights into the ecology of our forests. Most remarkable is the resiliency of trees and forests to recover following repeated, intensive and widespread disturbance. In the face of cutting, burning, windthrow, pathogens, broad-scale deforestation, and other novel impacts, forests have repeatedly regrown across the land.

Today's forests, however, are distinctly unlike those that existed at the time of European settlement or developed at various points since then. At the scale of an individual forest stand the trees are younger and more even-aged than in pre-settlement times. Also, the proportion of species has changed: the modern forest composition is more strongly dominated by sprouting and early successional species, whereas the long-lived and shade-tolerant species such as beech and hemlock are less common. Some species have disappeared. Chestnut, a tree that was historically important for wildlife as well as human economy, no longer occurs as a mature tree in our woods.

The overall structure of most New England forests has also changed; despite decades of regrowth, our woods are still missing the very large trees and the tremendous accumulations of standing dead and downed trees that characterize old-growth forests like the Pisgah tract, as shown in the pre-settlement diorama. Even sites that have been continually forested over the last 300 years — whose characteristics might be expected to remain fairly constant over time — have undergone a complete change in forest composition as a consequence of direct impacts like cutting, burning, and grazing and indirect human impacts like chestnut blight. There is little evidence of a tendency for these forests to return to their former characteristics.

Today, forest landscape patterns tend to change across ownership boundaries, reflecting the impact of land-use history more clearly than the gradual transitions in soils, topography, and drainage that determine natural patterns. This results in more local and abrupt variation in forest age, structure, and composition than in pre-European times.

At a broader regional scale our studies in central Massachusetts suggest that widespread land use has greatly homogenized the vegetation, leading to lower diversity and greater similarity in species composition in dissimilar habitats. In the pre-settlement forest, species abundance reflected different climatic and physiographic conditions. For example, there was more oak and hickory to the south and in the warmer Connecticut River Valley and more northern hardwood–hemlock forest in the northern, cooler uplands. Such differences have been largely obscured by the history of deforestation and forest regrowth over the past few centuries. Oak, red maple, black birch, and other sprouting species now prevail throughout the region.

One of the major lessons that emerges from the dioramas is that in order to understand our forests today we need to become deeply knowledgeable about their particular history. This historical perspective shows us that our forests have always been characterized by change and carry a strong cultural legacy of past human activity. This understanding should inform our predictions of our forests' future evolution, as well as our attempts to conserve and manage them.

Conservation Issues in the History of New England Forests

Richard Fisher and his colleagues at the Harvard Forest sought to apply their ecological and historical understanding of New England forests to the conservation and management of this landscape and its woodlands. Consequently, several of the dioramas in the Fisher Museum highlight conservation issues in the New England countryside.

A few of the models, such as the ones dealing with forest fires and their control, reflect regional and national concerns that peaked during the first half of the twentieth century or whose emphasis today has changed. Most of the models, however, are about issues and their interpretation that remain quite relevant to conservation, ecology, and forestry. For example, the large, beautiful double-width diorama depicting old-growth forest alongside Harvard Pond (diorama pp. 24–25) underscores the important lessons that come from observing natural ecosystems that are little disturbed by human activity. It also strikes a very modern chord: the current widespread interest in locating, studying, and protecting remnant old-growth forest stands across the eastern United States.

Similarly, the diorama depicting wildlife in the abandoned farm landscape (diorama p. 27) is directly relevant to our interpretation of the dynamics in modern animal populations and it furthers our understanding of the gradual decline of many of our favorite bird and mammals species that thrive in open agricultural grasslands and shrublands. Even a scene highlighting processes that are relatively uncommon in the modern forested landscape — like the one showing erosion in an intensively farmed landscape (diorama p. 29) — is nevertheless a valuable reminder of the major changes in soils and hydrology that occurred throughout the period of intensive deforestation in New England. Although these processes may operate very differently today, their past action on the land, resulting from historical land-use practices, is recorded as variations in the modern soil profile.

These dioramas remind us of the history of conservation issues in eastern North America, and also point up the continuity of certain conservation concerns through time. These perspectives are valuable to us as we seek to protect and preserve our modern landscape.

OLD GROWTH FORESTS

This small stand of old-growth forest on the shore of Harvard Pond, in the Tom Swamp tract of the Harvard Forest, survived the regional history of land-use and natural disturbance due to its sheltered position at the base of a rocky slope that was unsuited for agriculture. In their size, variety, and unmistakable antiquity, the overstory trees — visible here are hemlock, white pine, yellow birch, and beech — contrast strikingly with the young and moderate-aged forests now so common on abandoned farms and cut-over lands in the region. The pond and adjoining wetlands in the background support a diversity of aquatic and open-land plant and animal life that is relatively uncommon elsewhere in the largely forested landscape.

Such a natural and varied setting provides an ideal place to enjoy the beauty, wonder, and quiet of nature and refresh mind and body. This is what Professor Richard Fisher sought there; he appears in the model — anachronistically — in the company of Professor Nathaniel Shaler, a nineteenth-century geologist and Harvard dean who promoted the study of forests and to whose memory Shaler Hall at the Harvard Forest is dedicated.

Large trees blown down by occasional gales and now in various stages of decay are an important characteristic of old-growth forests. They testify to periods of relative stability punctuated by natural disturbance and ongoing biological processes of decay and regeneration.

The traces of even more ancient windfalls are shallow depressions, or "pits," and adjacent mounds in the soil, caused by the tearing out of roots no longer able to anchor the towering trunks.

Old dead snags remain standing from trees whose roots held but whose trunks decayed, and then snapped. Immense barkless snags, all that remains of veteran pines killed by lightning, provide food and habitat for insects and the animals that feed on them, as well as large cavities that offer shelter to owls, pileated woodpeckers, fishers, and bears. The whole forest floor, roughened by the moldering remains of forest debris, is spongy underfoot.

Surrounded by a largely human-dominated landscape, these ancient forests are an indispensable refuge for a wide range of organisms. To the ecologist, conservationist, and forester they also provide invaluable insights into the processes, dynamics, and structures that characterize natural forest ecosystems. Lessons from old-growth forests, in conjunction with a thorough understanding of the land-use history of the region, can be used to interpret, conserve, and manage our forests into the future.

The fate of this particular forest since the mid 1930s, when the diorama was constructed, underscores the dynamics of the natural world and old-growth forests. Like so many forests across the region, this old stand was severely damaged by the hurricane of 1938, but it is recovering and continues to grow undisturbed by humans. The contemporary forest differs in the details of its composition from its predecessor, but it has maintained the range of structures and features that characterize natural forests.

WILDLIFE HABITAT IN A DYNAMIC LANDSCAPE

New England's wildlife habitats and food resources have changed dramatically as the New England landscape has been transformed through time from largely woodland to an open mosaic of fields and woodlots and then back to forest. Understanding these landscape dynamics helps us interpret many of the major changes that have occurred in our wildlife populations, anticipate future changes, and adapt our management strategies accordingly.

Farm abandonment accidentally created many important new and dynamic habitats and favorable conditions for many species. In the first decades after a farm was abandoned, former agricultural landscapes were characterized by apple trees the farmer had planted, shrubs, seedlings, and diverse herbs (grasses and other nonwoody plants). These furnished fruit and browse readily within reach of many mammals, including deer and rabbits, and food and nesting sites for open-land bird species. Indeed, some species that were absent or scarce in the dense old-growth forests — such as the cottontail rabbit, woodcock, bobolink, and other grassland and shrubland animals — increased greatly as a result of the clearing and subsequent abandonment of farmland.

A few decades later, old-field white pine stands furnished good cover for mammals, and as the trees approached maturity they also provided food from herbs and the woody undergrowth of saplings and seedlings. When the stand was cut, the young hardwood sprouts that shot up in the intense light again provided ideal food for browsing animals like deer and rabbits.

Overgrown fields and young forests were not the only features of abandoned farms that changed the landscape and inadvertently provided new habitats. The old stone walls with their vines and shrubs furnished habitat and food for many animals such as chipmunks, mice, and weasels. Often the settlers dammed streams to obtain water power; the resulting millponds were used by a wide array of aquatic birds and mammals, including muskrats, otters, ducks, and herons. On the other hand, many wetland habitats were destroyed by early agriculture and widespread draining—a process that has continued across the northeast as a result of development practices, including the modification of the natural flowage of major streams and rivers by dams and channelization.

Forest management, by giving special consideration to wildlife, can create additional habitats or recreate many natural environments. The small, frequent silvicultural operations that are typical in a forest under multiple-use management (managed for a variety of purposes, not only timber production) can be planned so as to enhance the variety of food, cover, nesting, and rearing sites within the area of daily movement of the animals. In forests managed for intensive timber production, however, conflicts can arise. Deer can browse down the young trees that have seeded in or been planted to make a new stand, and beavers may cut down trees and flood woodland areas.

Forests have many potential uses — timber production, wildlife habitat, watershed protection, recreation, aesthetic amenity, grazing — but it is impossible to maximize all or most values and products from one forest area. The shift in our landscape to older, more continuously developing forests may encourage native woodland species but it also reduces habitat for open-land animals. Focusing on wildlife habitat as a management objective may reduce the value of the forest as a recreation amenity. As forest and conservation management proceeds, choices must be made as to which specific values to emphasize in specific areas. An understanding of history, biology, and management practices assists greatly in defining and reaching specific conservation and management goals, but ultimately the selection of those goals is a subjective process.

WILDLIFE IN A CHANGING LANDSCAPE

ACCELERATED EROSION WITH INTENSIVE LAND USE

In New England, as in many other locations, widespread land clearing and agriculture led to soil erosion. In the presettlement era, the thick forest cover prevented erosion by intercepting rain and binding the soil with a dense network of roots. Forests recycle more moisture to the atmosphere through evaporation and transpiration than does shrubland, grassland, or plowed fields, so they reduced the amount of water moving downslope into brooks, streams, and rivers. The sandy loam soils were highly permeable, so they easily absorbed water from precipitation, further reducing runoff.

This tight control over hydrology and water movement changed with land clearing for agriculture. Throughout the historical period of deforestation and intensive agriculture, soil moved on a local scale, across a field, and on a broader scale, down hillslopes and through major river systems. This movement of earth impaired agricultural productivity at the time and it continues to influence our soils and landscape today. (Importantly, however, erosion resulting from agricultural use is less marked across much of upland New England where the glaciers had already done the job of scraping much of the topsoil than in the unglaciated regions farther south with deeper soils.) Today, soil erosion is uncommon across the New England uplands because once again this countryside has a thick forest cover.

Many factors contributed to the increase in erosion in the agrarian landscape of the eighteenth and nineteenth centuries. The removal of forest cover increased the direct impact of rain on the soil surface and allowed an increase in surface runoff. Bare plowed fields lacked the binding network of roots and so were highly susceptible to erosion. They were made more vulnerable by the practice of plowing downhill, which was easier for the oxen and horses and probably also for the men who drove them. The construction of roads that cut into hillslopes destabilized soils and interrupted the natural subsurface flow of moisture. In addition, constant disturbance by the hoofs of grazing animals churned the surface deeply in swales and barnyards, and carved trails along the contours of slopes, which acted almost like roadcuts.

Each of these processes accelerated the natural redistribution of soil material downslope. The diorama shows numerous examples of accelerated erosion in a broad valley flanked by glacial terraces of silt and fine sand, which are unusually vulnerable. Gullying has been caused by the concentrated water flow on a cleared terrace, which has focused runoff in one larger channel. Sheet erosion, erosion over a broad area, has occurred across a field that has been plowed downslope. Mass movement of soil has taken place along the roads and where the terrace is undercut by a stream. Meanwhile, the deposition of fine materials across the meadows adjacent to the river, which is a natural consequence of flooding due to spring snow melt has been greatly accentuated by the larger concentrated pulse of snow melt, resulting from reduced forest cover. The diorama also shows the ability of wooded hillsides to resist erosion.

A careful observer will easily note the signs of this earlier soil movement in today's landscape: Many knolls and upper slopes are missing the top layer of the organic and nutrient-rich topsoil that provides much of the productivity in forests. This missing material has been relocated to form the deeper soils on more level terrain downslope. The observer can see where much of it ended up: in deeper, stone-free soils on the upslope side of stonewalls, which served as dams to the eroding fine materials; in fanlike deltas of fine soil extending outward from small streams into ponds and lakes; in more substantial debris fans at the base of long steep slopes. And much of the soil may have been deposited on the floodplains of larger rivers or washed out to sea.

The consequences of this historical episode of soil movement for today's forests have been inadequately studied and are poorly understood. Undoubtedly, however, the legacy of erosion is a permanent feature of many New England landscapes and affects the distribution and growth of plant and animal species.

EROSION WITH INTENSIVE LAND USE

The diorama shows a wild fire that most likely was ignited by a camper or passerby. The fire tower is typical of those built by state and federal agencies throughout the northeastern United States in the early 1900s, motivated by a widespread concern over the detection and control of fire. Such towers, like the one atop Prospect Hill in the Harvard Forest, continue to be staffed during periods of high-risk weather when drought conditions prevail. The scene underscores not only the fact that fire is a potentially important force shaping New England forests, but also that understanding the historical and modern role of fire is a subject of great interest to ecologists, conservationists, and foresters.

In general the rate of naturally occurring forest fires is low in New England. In the conifer-dominated forests of the western United States and the northern boreal forest region, forest fires are often started by lightning and can spread rapidly as "crown fires," which can cover hundreds or thousands of square miles. In most of New England, by contrast, the broadleaf foliage and the abundant rain that accompanies lightning in intense thunderstorms make natural fires uncommon.

The ecology of fire across New England is quite varied, however. In northern Maine, like other areas of coniferous northern boreal forest, the relatively dry, resin-containing foliage of conifers is more flammable than the broadleaf vegetation across the rest of the region. Most of New England is characterized by broadleaf deciduous forest, and once the forest is fully leafed out during the summer, it is relatively non-flammable because of the plants' high moisture content and the high humidity of the understory. Consequently, most fires occur in the spring or fall, when winds can dry the understory and fan the fires. These fires are "surface fires," which run along the ground, consuming leaf litter, dead woody debris, and organic soil layers and injuring or killing thin-barked trees through scorching. An intense fire can exert a long-lasting impact on forest structure and composition:

many herbs, shrubs, and hardwood trees sprout prolifically after fire, whereas species like white pine and hemlock, which do not sprout, must reestablish by seed and tend to be suppressed by fire.

Despite the region's low-risk of natural fire, human-set fires have probably played a considerable role in New England from the pre-settlement era down to the present. Ethnographic evidence that Native Americans made widespread use of fire to clear forests, improve agriculture, and facilitate hunting has led to the widely held notion that fire was an important factor influencing local forest landscape patterns for millennia before European settlement. Solid evidence to back this assertion is scanty and additional research on this subject is needed. However, it is safe to conclude that in pre-settlement New England the role of fire — both set and naturally occurring — was insignificant in the higher-elevation northern hardwood region, but greater in the southern, coastal, and riverine locations, where Native Americans were concentrated and where relatively dry oak and pine forests prevailed.

Following European settlement, during the period of deforestation, agriculture, abandonment of farmland, and forest regrowth, the frequency and intensity of fire increased in much of the eastern United States. Fires were started by careless farmers, loggers, campers, sparks from wood- and coal-burning trains, and even the glowing wadding that was ejected from the barrels of primitive firearms. This increase in fire activity had an immediate effect on forest composition that has passed down to our modern forests: the abundance of sprouting hardwoods, like oak, and early successional species like paper birch increased, whereas fire-sensitive species like hemlock and beech were selectively removed.

Major wild fires in the late nineteenth and early twentieth centuries, often exacerbated by slash left from poor logging operations, also shaped national policy in forest management and conservation, for they led to a widespread concern with forest fire

and a major drive for fire suppression. Throughout much of the twentieth century fire has been perceived as strongly detrimental to forests and wildlife and as the cause of a widespread decline in site fertility, forest productivity, and timber quality.

There is still much to be learned about the history and past role of fire in the New England landscape as well as its potential use as a management tool in shaping forest and other ecosystems. Although we understand much about its ecological impacts, we remain challenged to decipher its past frequency and geographical distribution, its variation through time, and the effects that this history continues to exert on our modern landscape.

Concern by early conservationists over the apparent destructive impacts of fire on forest ecosystems led to a very successful national effort to reduce ignitions and enhance detection and control of both human-set and natural forest fires throughout much of the twentieth century. In New England this all-out effort was accomplished through a variety of measures. Throughout the thirties and forties most of the fire towers that are still visible in state and national forests were constructed. Local fire departments were trained and equipped to fight wild fires. Across the landscape a system of waterholes was created, and a dense network of roads was developed in many remote areas and on most public lands to increase access for fire-control equipment.

The forest fire diorama (preceding page) shows a typical fire-fighting scene from the 1930s, in the era of active suppression of all fires. It shows firefighters fighting the blaze with backpack pumps. Water is delivered to the fire from local ponds or water-holes by gasoline pumps. The fire tower is staffed by a watcher equipped with telephone or radio communication. The diorama to the right depicts the site after the fire has been put out.

Over the past few decades, ecologists, foresters, and conservationists have reconsidered the wisdom of the all-out effort to eliminate fire from our landscape. In many parts of the world fire is actually an important natural process, one that over millennia or longer has shaped a wide variety of ecosystems, from grasslands to shrublands to many types of forest. The generally successful effort to reduce fire in most regions of the United States has generated many pronounced changes in vegetation, ecosystem function, and wildlife populations. In some areas, particularly in the dry West, fire prevention activities may prove ineffective or even disastrous in the long run, as the amount of fuel accumulates to abnormally high levels during the fire-free period. Then, when there eventually is a fire, it burns with unusual intensity. In other regions, like New England, fire may

have been an important cultural factor in both the Indian and European periods that has shaped the landscape characteristics we see around us today. Without the ongoing occurrence of fire and other human-induced disturbances the vegetation and landscape may change quite rapidly and populations of valued species that depend on the open conditions and specific structures created by fire may decline.

The recognition of fire's historical importance has led to an increased appreciation of its value as a management tool to shape vegetation in directions that are desirable for ecological and also historical reasons. Some creative new approaches for using fire to manage natural and planted vegetation are currently being applied. In remote parts of the United States — for example, in many wilderness areas and national parks — under certain conditions some naturally ignited fires have been allowed to burn under close supervision as part of the long-term management plan for these regions. A more proactive use of fire is "prescribed burning," when a fire is purposefully set and closely monitored by knowledgeable experts under highly restricted meteorological and fuel conditions in order to achieve a specific management goal. Some typical reasons for igniting a prescribed fire are to reduce fuel loading of litter and woody debris; to open the understory or overstory and promote particular plant species; or to shape and create specific wildlife habitats. Controlled fires have been used to a limited extent in all New England states, including Massachusetts, especially in pine and oak forests, sand-plain ecosystems, heathlands, and grasslands. In shrubby and grassy areas, fire is often used to restrict the growth and expansion of trees and other woody vegetation that would otherwise rapidly change the habitat.

Although fire suppression, especially near dwellings, remains a major concern in many of our forest areas, we now turn to fire-fighting tools and information on fire behavior to manage some New England landscape purposefully with fire.

FOREST FIRE MANAGEMENT

Forest Management in Central New England

By coupling the forest history of New England to an understanding of the ecology of the region, the biology of forest trees, and society's demands for natural resources, Richard Fisher developed a comprehensive approach to forest management that he and his students came to call "ecological forestry." Because this silvicultural approach was based on the study of natural stands and native species and attempted to work with the basic biology of forests in their natural landscape setting, it provides a clear precursor to the "new forestry" and "ecosystem management" approaches that have emerged in the late twentieth century. The following dioramas present a range of examples from this system with an emphasis on situations that were common in the New England landscape in the early twentieth century. Although they obviously do not include the entire range of recent advancements in forestry techniques, equipment, or products, they offer great insights into approaches for reading a forest stand and managing on a local scale. Thus, they remain important instructional tools that offer insights to individuals interested in forest management and in particular to the owner of small woodland areas.

The dioramas in this "silvicultural" series depict both hardwood and softwood management, but emphasize softwood management because pines were the dominant and most valuable trees when Professor Fisher and Harvard Forest scientists were developing these ideas. Hardwoods had been in high demand for fuelwood since the early nineteenth century and had been heavily and repeatedly cut. Consequently, they had not generally grown to large size or high quality in the 1930s, and local markets for hardwood sawtimber were less well developed.

Today, hardwoods, especially red oak, are among our most valuable species, as stands that regenerated after the cutting of old-field white pines in the early 1900s or the salvaging of hurricane blowdowns in 1938 are maturing. Although markets shift with the availability of a quality resource and forest structure and composition change through time, the best forest management still combines a thorough understanding of site history, site quality and anticipated markets. From the early 1900s into the 1930s a variety of conifers were planted in plantations at Harvard Forest to study their growth, interaction in various mixtures, and ultimate productivity. Similar plantings were made across New England, especially on state forest land and around municipal reservoirs and watersheds. As these plantations mature many are being harvested and then left to regenerate to natural forest types.

As a consequence of the abundance of forests and the often marginal economic returns from forestry in New England, many of the intensive practices introduced by Fisher and others early in the century, like replanting after cutting, early thinning of stands, or pruning of lower branches are not practiced on a widespread basis.

EARLY TREATMENT OF A HARDWOOD STAND

This diorama shows two stages in the development of a hardwood stand after the cutting of an old-field pine forest. In the background on the left is the edge of a 60-year-old white pine stand about to be clear-cut for lumber. Just prior to logging, seedlings and small saplings of hardwood species in the understory are cut off close to the ground. Such undergrowth (called "advanced growth"), if left uncut, will be bent or broken by falling trees. Even if it were possible to save the stems following this damage, they would not develop into straight, "thrifty" trees. After cutting, the very small stumps send up well-formed sprouts that will grow rapidly and will be straight enough to form good timber. Therefore, this cutting will actually ensure better young crop trees in the future. These sprouts may grow as much as six feet per year, if they are not browsed back by deer or moose.

In the center foreground is an area where the mature pines were logged six years earlier. Sprouts from the stumps of the hardwood undergrowth have been naturally supplemented by seedlings, many of which are early successional or pioneer species with readily dispersed seeds such as gray and white birch, red maple, and cherry. Workers are cutting away the smaller stems among the multiple-stemmed hardwood sprouts and are removing the short-lived pioneer trees wherever they

dominate and threaten to crowd out more desirable trees. Thus the woodland owner or forest manager selects the most desirable trees for the future stand at an early age, concentrating and speeding up the wood production in the stand on a smaller number of the best trees. Naturally, the selection of desirable stems is a somewhat subjective choice that may be guided by the economics of wood production, aesthetics, or specific conservation goals, including the promotion of particular species and habitats. Today, such weeding is economical only on small, individual woodlands or in association with other forestry activity that can pay for this labor-intensive work.

A second weeding at 15 years may further speed the development of a healthy stand of vigorous trees and enhance the process of concentrating the tree growth and production of wood on the most desirable stems. At this point many of the smaller, shorter stems can be left, because they actually serve a useful purpose: these so-called "trainers" shade the lower trunks of the dominant trees, thus suppressing the growth of lower branches on them, which form undesirable knots in the sawtimber. These smaller trainers help the crop trees to form straight trunks and clear, knotless wood.

EARLY TREATMENT OF A HARDWOOD STAND

IMPROVEMENT CUTTING IN A HARDWOOD STAND

Hardwood stands that have not been thinned or weeded in the first several decades of forest development contain trees of many kinds, shapes and sizes: some are straight and sound and of valuable species; others are forked or crooked sprouts from large stumps; and others are of short-lived species that will soon begin to die. Like weeding, improvement cutting, which is done at a later stage in stand development, generally aims to favor the best-formed and healthiest trees for the future by removing inferior stems that overtop or unduly crowd the selected better trees. The definition of "inferior" and "better" depends on the objectives of the landowner. Somewhat different criteria may be employed for timber production, aesthetics, or wildlife management. For example, paper birch is often left for its attractive appearance, whereas red oak is favored as a valuable timber tree and oaks and beech are valued as important sources of "mast," or nuts that are food for many wildlife species. At this stage most of the trees are big enough to be used for fuelwood or pulp, so improvement cutting often provides a valuable product.

The portion of the stand on the left has not yet been treated, but a forester has marked the trees that are to be removed. He has just marked a vigorous overstory aspen because it is short-lived and will soon begin to die. Cutting it now will generate a useful product and will free up substantial space for surrounding trees.

In the lower center the improvement cutting is in progress, resulting in piles of logs. Gray birch, aspen, and poorly formed multiple-stemmed stump sprouts of many species, including red maple, are being cut to favor well-formed red oak, paper birch, and white ash. At this age many of the red oaks are beginning to outgrow and overtop the other species; some of the smaller trees are being left as "trainers," which will shade and eventually kill the lower branches of the "crop" trees, thus increasing their value as knot-free sawtimber.

Foresters call a tree like the tall one in the center-left background a "wolf" tree: its size and wide-spreading branches indicate that it has grown under open conditions in the past. A remnant from an earlier stand, this tree was left standing because it was deemed unsound or of too little value to be worth harvesting. Now it must be eliminated so that it will not compete with the growing stand. It has been "girdled" — the bark and cambium have been cut through all around the trunk to interrupt the flow of water and nutrients — but left standing so that it will fall apart slowly and will not damage the surrounding trees as it falls. Over the next decade or so it will provide nest sites, perches, dens, and food for wildlife.

In New England it is typical for a substantial variation in species composition to occur with slight variation in elevation and landscape position. For example, on the right is a hardwood swale, a slight lowland that has much wetter soils than the surrounding uplands. There, foresters will free the straight, single-stemmed white ash and yellow birch from intense competition by cutting red maple stump sprouts. Such local variations demand that all management activity be undertaken with careful attention to site and forest condition as well as history.

IMPROVEMENT CUTTING IN A HARDWOOD STAND

The improvement-cutting stage pictured in the previous diorama is eventually followed by thinning, which continues the process of removing trees that compete with selected crop trees. An overcrowded stand is likely to be full of tall spindly, slow-growing trees, unless some trees that compete for water and nutrients in the soil and for crown space overhead are removed. The crowns of the selected trees need room to expand, and this in turn will speed up the whole tree's growth. Large crowns with plenty of foliage exposed to the sun are needed to produce enough wood to thicken the main stem (also called the "bole") quickly.

Under natural conditions, despite overcrowding some trees will gradually surpass the others in height and will begin to suppress their neighbors and dominate the stand. Thus the overall density of stems in the forest will decline with age. Thinning speeds up this natural process and also ensures that the selected, most desirable trees become dominant. In central New England, where red oak is present it generally starts to outgrow other species after two or three decades, but paper birch can keep up with the oak for 60 years or more. This process can be seen in many mixed hardwood stands across the region if one looks upward to compare the size and height of individual red oak crowns to those of the neighboring maples and birches.

The thinning taking place in the hardwood stand to the right of the road is a "crown thinning" or "high thinning," whose purpose is to give the best oak, birch, and white ash plenty of crown space by cutting the poorest of the competing overstory trees. Not all smaller trees are cut. Great care is taken to save some overtopped smaller trees as trainers to shade the crop tree stems and help "prune off" their lower limbs. Wherever possible the trees to be removed are those that have noticeable imperfections such as crooked or multiple trunks or a broken crown, or that are long slim "whips" that abrade against the crop tree crowns.

At the left center the thinning has been completed — the logs dressed and neatly stacked; in the center thinning is in progress; and on the right a forester is marking the trees to be removed with spots of red paint from a paint spray gun. The portion where thinning has been completed contains but two stand elements: overstory crop trees and smaller trainers. Prominent among the former are red oak, paper birch, white ash, and black birch.

The white pine stand just visible on the left side of the road has been subjected to a "low thinning": all smaller trees are removed. In pine stands, overtopped trees are not as effective as trainers as in hardwood stands because the white pine retain their dead limbs for a very long time.

FIRST THINNING, WHITE PINE AND HARDWOODS

This is the same stand as the one shown in the previous diorama, approximately 25 years later. Many of the largest trees are now 12 to 16 inches in diameter and of fine quality for timber. This excellence in growth and form for timber production is due to the weedings and thinnings applied periodically since the stand was quite young. Untreated stands do not yield timber of the same quality.

The thicker, lower parts of the boles (trunks) of trees removed in this thinning are suitable for sawing into lumber or for producing veneer, while the smaller upper portions and the larger branches are useful as firewood or pulp. As in all previous treatments, the aim is the continued improvement of the stand by removing the poorer elements at such times and in such amounts as will provide the remaining favored trees with optimum conditions for growth and development.

This third thinning reduces the number of timber trees in the hardwood portion of the stand to about 100 per acre, spaced about 20 feet apart. Only the best individuals of the most desirable species, such as red oak, white ash, paper birch, and sugar maple, are left to form the final stand.

The presence of the trainer trees cleared the boles of the timber or crop trees of dead branches up to about 32 feet, or two full 16-foot log lengths by the time the stand was about 30 years old. During the remainder of their lives these trees will produce clear, straight-grained wood, capable of meeting the most exacting requirements of the market.

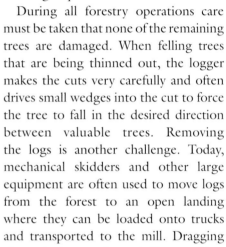

During all forestry operations care must be taken that none of the remaining trees are damaged. When felling trees that are being thinned out, the logger makes the cuts very carefully and often drives small wedges into the cut to force the tree to fall in the desired direction between valuable trees. Removing the logs is another challenge. Today, mechanical skidders and other large equipment are often used to move logs from the forest to an open landing where they can be loaded onto trucks and transported to the mill. Dragging 30-to-60-foot-long logs through the woods can inflict considerable damage on remaining trees, forest soils, streams, and wetlands unless great care is taken. One way to shield the remaining trees from scrapes is to leave a few undesirable trees that function as "bumpers" while the logs are being maneuvered through the forest. These bumpers are cut last, leaving only the best crop trees in the stand. Timing is also important. Foresters can minimize soil damage by logging when soils are dry or frozen, and must be careful to plan access trails and stream crossings in ways that reduce severe impacts on fragile wetlands and waterways.

THIRD THINNING, WHITE PINE AND HARDWOODS

CONVERSION OF CORDWOOD TO FUTURE SAWTIMBER

On dry upland sites such as the one pictured, hardwoods often do not grow vigorously and are likely to develop crooked stems, like those in the stand on the far right. Pines, on the other hand, can grow quite well and produce high-quality timber on such sites, provided they are not crowded or overtopped by the hardwoods.

With some effort, a hardwood stand like the one at the right can be converted to pine or another conifer species. In the center, the hardwood forest has been cut for cordwood and the slash burned, and now two crews are at work planting conifers. They are using transplants, small saplings that were first grown in a nursery for two to three years, to ensure better initial growth rates. The hardwoods will sprout after cutting and may temporarily crowd the sides of the planted conifers, thus helping to prune off lower branches. However, the sprouts must be watched and controlled through additional cutting so that they don't overtop the pine. On dry sites, where the hardwoods do not resprout vigorously, few weedings are necessary to keep the conifers from being shaded out. On moist sites, the greater vigor of hardwoods calls for frequent weeding, which is usually not economically feasible, and these sites are generally best managed for hardwood forest.

Historically, many different species have been available for planting and the ones chosen have depended upon such factors as suitability to a given site, susceptibility of the seedlings to insect or fungus attack, the influence of one species upon another if they are to be grown together, and the desired product. Conifer species that were planted on the Harvard Forest and on many state forest lands in central New England include white and red pine, white and Norway spruce, and European and Japanese larch. Today, such labor-intensive tree planting is not cost-effective on most sites in central New England and is seldom undertaken. The major exception would be Christmas tree farms, which produce a fairly valuable product in 15 years or less. Such trees do, however, require additional maintenance and care, including mowing, pruning, and shaping.

CONVERSION OF CORDWOOD TO FUTURE SAWTIMBER

INCREASING WHITE PINE IN HARDWOOD

Pines are more competitive with hardwoods in areas of dry, sandy soil than on rich, moist sites. On the former it may be advantageous — and possible — to supplement the hardwood trees with groups of white pine. At the time the dioramas were developed there was a strong emphasis on managing forests for pine production because pines had the greatest value in the local market. Today hardwoods are more valuable and it is unlikely that the efforts described below would be made to convert a hardwood stand to pine.

The diorama shows various approaches to such conversion. Two of them also represent attempts to control predation on white pine seedlings and saplings by two destructive native beetles, the pales weevil and white pine weevil. On the left-hand side, a cluster of white pine seedlings is being planted on a small knoll. A few years previously a white pine forest was logged, and branches and windrows of logging slash were burned. After such logging, a delay of at least two years before planting is necessary because the pine stumps and fresh pine cuttings attract the pales weevil, which then also feed on any small seedlings on the site. Young pines growing in the open are most attractive to this insect. The hot fires of the burning slash destroyed many of the hardwood saplings or "advanced growth," reducing hardwood competition for the white pine.

In the center is a young stand of groups of white pine mixed with hardwoods. In this case the pine was established through what is termed the "shelterwood system of regeneration" in which a significant proportion of the overstory is left to provide seeds and shelter for the regenerating stand and is then later removed. One aim of this system is to reduce damage from the white pine weevil. This beetle frequently kills the tip of the main shoot of pine seedlings, causing many side branches to begin to grow into vertical stems. This leads to the distinctively contorted multistemmed pines that are common across New England. The shoots of young white pine are less likely to be attacked by the weevil in the partial shade beneath a shelterwood cut than in completely open conditions.

Consequently, in this method pine seedlings were allowed to become established and grow for a while under the shelter of the previous pine stand before it was logged. To encourage this, some mature trees were cut, and the seedbed was improved by scouring and exposing the mineral soil beneath the litter layer as the logs were dragged out. Growing conditions are also improved by admitting increased light to the understory. Several years later, when the remaining overstory trees are harvested, numerous white pine saplings and hardwood saplings from both seeds and sprouts are growing on the site. The fast-growing trees of inferior timber species and even faster-growing stump sprouts are being cut to free the best future trees of both pine and hardwoods.

Barely visible at the right is a site that supported a pine forest that was cut seven years ago, in a good pine seed year. Some small groups of pine became established along with the hardwoods, and the best pine and hardwood are now being released from competition by cutting the overtopping trees of less desirable form and species.

In all three cases forest management is seeking to establish a forest mosaic with groups of white pine and groups of hardwoods. These distinct groups create a diverse structural pattern in the landscape, which enables the pines to grow more effectively than they would if completely intermixed with the rapidly growing hardwoods.

INCREASING WHITE PINE IN HARDWOOD STANDS

RELEASE OF PINE FROM SUPPRESSION BY GRAY BIRCH

Some abandoned fields or other open sites seed in heavily with both gray birch and white pine; most often this occurs on dry sites with an abundant supply of birch seed trees and not many white pine, especially if the site has been burned, removing organic litter and exposing bare soil. Although much shorter-lived than the pine, gray birch grows much more rapidly at first and soon completely overtops the pine. Though the light shade created by the birch slows the growth of the pine, it provides some benefit to the pine by making its shoots less attractive to white pine weevils, so damage from this insect is reduced. Eventually, the white pines should be released from competition by cutting the birches when they start to whip against and damage the desired pines.

If the white pines are not released, white pine mortality under the birch depends to a great extent on the fertility of the soil. On moist soils the birch may grow vigorously and cast dense shade, and the pine will be almost completely eliminated by the time the stand is 20 to 30 years of age. On dry, sandy soils, which conifers can tolerate better than hardwoods and where the birch may grow weakly, the pine may persist until the gray birch reaches maturity and dies. Thereafter, the pine is able to grow and develop into a nearly pure stand.

Studies by Richard Fisher, Albert Cline, and others at the Harvard Forest have shown that there is an optimal height for releasing the white pine to ensure that it will not again be overtopped by the sprouts from the birch stumps. For moist soils, this point is reached when the pine is 15 feet high. However, at this time the birch may not have attained a suitable size for cordwood. All factors considered, the release cutting should be made at the earliest age at which the birch is marketable, ordinarily between 18 and 25 years of age.

RELEASE OF PINE FROM SUPPRESSION BY GRAY BIRCH

PRUNING WHITE PINE TO PRODUCE BETTER LOGS

One characteristic of white pine in pure stands is the persistence of dead branches that cause loose knots in the resulting lumber. If high-quality, knot-free lumber is desired, the lower branches must be sawed off close to the live stem when the trees are young. Such pruning also produces a more aesthetic appearance in a highly visible stand.

Naturally seeded dense stands on old fields are especially well adapted to pruning, because close spacing shades and kills the lower branches while the trees are small and forces them to grow straight. On the far left side of the model a young stand of such origin is now receiving its first pruning. The dominant trees are about 17 feet high and are pruned up as high as a person can reach, about 6 or 8 feet. From an economic viewpoint it is important not to expend effort pruning trees that will not be final crop trees, and so approximately 100 to 150 of the best-formed "taller trees" are identified and pruned. These will form the final high-quality crop.

In the center a similar stand has reached the age (about 30 years) when the final pruning is in order. Using a ladder or a long-handled saw (called a pole saw), a forester clears away branches 16 feet up the trunk (a standard log length), or less if a

crook in the bole occurs at a slightly lower point. Neighboring trees that interfere with the growth of the pruned crop trees should now be cut or girdled to provide more room for the top branches and crown of the crop tree to expand.

On the right is a plantation of widely spaced trees in which the dominant trees became forked and crooked as a result of earlier attack by the white pine weevil. Here the crop trees are chiefly ones that were partially overtopped and shaded and therefore were less subject to attack, and less able to become bushy if attacked. To release such trees, it was necessary at the time of the first pruning — about 20 years ago — to girdle neighboring, severely weeviled "dominants." Most of the trees girdled earlier are now dead, and a workman is girdling others that threaten to suppress the pruned trees.

Girdling (or selective herbiciding) of trees is often an easy and cost-effective means of thinning a stand when there is no motivation to remove the resulting wood. Under these circumstances girdling is preferable to cutting for a number of reasons: Dead trees left standing protect neighboring pruned crop trees against sun scald and snow breakage. Also, the dead stems eventually rot and do less damage to the surrounding trees when they fall.

PRUNING WHITE PINE TO PRODUCE BETTER LOGS

GROUP SELECTION METHOD OF HARVESTING WHITE PINE

The local landscape typically contains different soils and habitats in close proximity as a consequence of the interaction between the bedrock geology and action of the last glaciation. On dry sites such as glacial outwash deposits of sand and gravel or gravelly kame deposits that form terraces along hillsides, white pine frequently forms nearly pure stands. Hardwoods require moister and more fertile soil for vigorous growth, and in their absence the pine can maintain itself as a semi-permanent forest type. Generally, such areas contain a mosaic of small, even-aged patches or groups, each of which established naturally after some disturbance, such as fire, windstorm, or cutting or other land-use activity.

Clear-cutting large areas of white pine growing under such conditions is not only wasteful of the smaller trees, which are felled before reaching the optimum size, but undesirable from the standpoint of the next crop of trees. Full exposure of the soil allows dense mats of blueberry, wintergreen, and other ground plants to develop and these, in turn, seriously hinder reproduction of pine by seed. (Such cutting may, to be sure, produce desirable improvements in wildlife habitat.) Restocking by means of planting is expensive and almost equally unsatisfactory. But where there are approximately even-aged small stands in proximity to other somewhat different-aged stands, a partial cutting system known as "group selection" is possible. This approach is based on the ability of a stand to naturally reseed a nearby area.

On the left (see detail) a mature group of trees, about 60 years old, is being clear-cut. This small cleared area will be well seeded by the oldest neighboring groups, such as the one shown in the center background, which is 50 years of age. The still younger group on the right margin, about 30 years old, was seeded by a nearby stand; it will not be ready to cut for several decades. In the center foreground is the very youngest age class: seedlings that grew up after a group cutting 10 years earlier. Pine seedlings are most abundant on two moss beds, one near the farthermost figure, and one growing on rotting slash directly in front of the individual in the foreground. Pine seedlings occur in a more scattered pattern on less favorable seedbeds.

At the right center is a bed of reindeer "moss" (actually *Cladonia* lichen) in a natural opening. Very little pine has been able to establish itself on this drier, less favorable substrate, except under the protection of a clump of gray birch "nurse" trees, which offer shade and somewhat moister conditions. In the background at the left center between the patches of tall pine is a clump of black or red spruces and larch growing in a small boggy depression in a rolling landscape characteristic of morainal deposits left by the glacier.

NATURAL REPRODUCTION, GROUP SELECTION

The shelterwood method is a silvicultural approach in which a new generation of trees is established naturally under the shelter of the old trees by a series of partial cuttings intended to stimulate seed production, create favorable seedbed conditions on the forest floor, and supply the young seedlings with sufficient room and light for healthy growth. This natural regeneration is both much less expensive than planting seedlings and more likely to succeed without continuing frequent treatments. Reproducing white pine by shelterwood regeneration is also advantageous as it can help prevent damage from the white pine weevil. On open full-sunlight sites, white pine saplings are more vulnerable to the weevil, which attacks their leading shoots. When the young white pines grow in the partial shade of the shelterwood, the terminal shoots are less often attacked by the weevil.

Two shelterwood methods — the uniform and strip methods — are shown in the diorama. On the left the uniform shelterwood method has been applied to a mixed stand of white pine, oak, and hickory growing on dry soil. Over the previous years, two partial cuttings of all three species opened the crown canopy of the parent stand uniformly; the trees that remain contain dominant wide crowns that should produce large quantities of seed. Under the conditions of increased light, seedling reproduction has become established rather uniformly. In the final cutting of the old stand, a few years hence, the remainder of the overstory trees will be removed, releasing the new stand to grow under very open, high-light conditions.

On the right the strip shelterwood method of regenerating white pine is shown. Removal of the parent stand will progress by strips from left to right, with cuttings spaced several years apart. All of the large trees have just been removed from the oldest strip of pine, the one along the road. The strip needs to be wide enough to provide open conditions for reproduction and growth, but narrow enough to ensure that it receives abundant seed and some shade from the adjoining forest stands. On the adjoining strips, partial cuttings are being made to increase the growth of the small seedlings already present and to encourage the establishment of more seedlings.

NATURAL REPRODUCTION, SHELTERWOOD

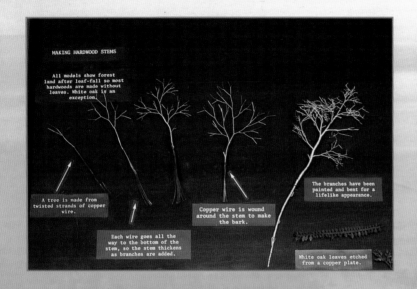

MAKING HARDWOOD STEMS

All models show forest land after leaf-fall so most hardwoods are made without leaves. White oak is an exception.

A tree is made from twisted strands of copper wire.

Each wire goes all the way to the bottom of the stem, so the stem thickens as branches are added.

Copper wire is wound around the stem to make the bark.

The branches have been painted and bent for a lifelike appearance.

White oak leaves etched from a copper plate.

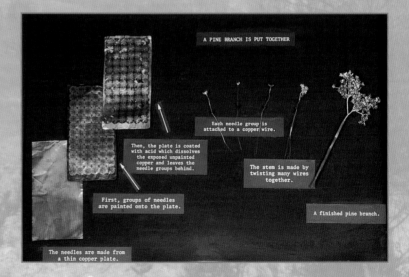

A PINE BRANCH IS PUT TOGETHER

Each needle group is attached to a copper wire.

Then, the plate is coated with acid which dissolves the exposed unpainted copper and leaves the needle groups behind.

The stem is made by twisting many wires together.

First, groups of needles are painted onto the plate.

A finished pine branch.

The needles are made from a thin copper plate.

Artistry and Construction of the Dioramas

The artisans of the Guernsey and Pitman studios were justifiably proud of their creations and consequently they produced a final diorama documenting the techniques that they had developed to make the remarkably realistic models. This diorama consists of four sections illustrating the different stages in the construction process, from the initial model framework on the left through to the completed model on the right (on page 59).

On the far left we see the base built of a framework of battens with copper screening stretched across them. Toward the rear the first coat of surfacing material, gesso, has been spread. The curved background is bare sheet metal that has been thoroughly cleaned. A "human figure," just out of view in the photograph is merely a wire frame.

In the left center the undulations of the terrain are built up of excelsior dipped in gesso with a second coat of gesso applied. A stone wall of the same material has been started. The background has a first undercoat of paint. The individual has been bent into the desired pose, and built up in wax to the correct proportions. The trees are first shown completely built, but in the copper stage. They are made of copper wires twisted together to form the branches and then the long ends are bound together by wrapping wire about them to form the smooth trunk (details at left). This technique produces the very realistic tapering effect as individual twigs formed from one wire join, progressively, to form minor and then major branches, and ultimately the central stem or bole. The leaves (where present) and needles are etched from sheet copper and soldered on to copper wires, which in turn are twisted into branches. This technique developed by Guernsey and Pitman specifically for these models closely mimics the structural development of actual trees, and is a major factor contributing to the realism of the models. In fact, careful attention by the artists to tree-growth form allows most of the trees to be identified to species by their branching pattern even without leaves.

In the right center, the base has a second coat of paint, and some larger stones and sticks provide roughness. The stone wall is complete and the figure is completed and ready for painting. The background has a second base coat and the picture has been sketched out. The trees have been dipped in molten solder to stiffen them and a coat of surface paint has been applied.

On the far right the model is completed. Small sticks and stones have been added, grass has been inserted with moss to remove any hard look, and the base has been painted. The figure is painted and finished. The trees have been sprayed with surfacing paint and oil color. They are then painted by hand to give each one the identifying bark variations of the individual species. The foliage is sprayed with paint, then the branches are touched up and color variations in the foliage are done by hand. The background is painted with particular attention so it blends in with the foreground. This blending, combined with the curved background and the use of forced perspective, enables the models to provide a sense of great depth within a relatively shallow space and makes it difficult to detect where the three-dimensional model becomes a two-dimensional backdrop.

The attention to realistic detail extended well beyond the actual crafting of each representative scene as the artists blended distinctive and varying cloud patterns with a range of whimsical

details, including diverse wildlife species and humans in a variety of poses and engaged in numerous activities. The variety and complexity of these natural and cultural elements yield new observations and insights with each viewing of this remarkable collection of models.

Construction of the dioramas was begun in 1931 in the Harvard Square studios of Guernsey and Pitman and they were completed in 1941 by Theodore Pitman and his associates with the philanthropic support of Dr. Ernest G. Stillman. The majority of the dioramas were designed by Richard T. Fisher, the first director of the Harvard Forest, to whose memory the Fisher Museum is dedicated, and Albert C. Cline, who became the Forest's third director. The wildlife management model was developed by Neil W. Hosley, instructor in wildlife management at the Forest; the soil erosion model was developed by Kirk Bryan of the Harvard Department of Geology and P. Rupert Gast of the Harvard Forest staff. The forest fire models were constructed under the supervision of M. C. Hutchins, State Fire Warden, and J. P. Crowe, Supervising Fire Warden, of Massachusetts.

DESIGN AND ARTISTRY OF THE DIORAMAS

SUGGESTED FURTHER READING

The dioramas provide a marvelous introduction to the history, ecology, and management of New England forests. The following sources expand on many aspects of forest ecology and review related research developments and topics.

Allport, Susan. *Sermons in Stone: The Stone Walls of New England and New York.* New York: Norton, 1990.

A nice introduction to the range of types and uses of stone walls and related cultural structures in the New England landscape.

Anagnostakis, Sandra. L. "Chestnuts in Our Forest," *Connecticut Woodlands,* Summer, on-line edition, June, 1998 (published by the Connecticut Agricultural Station, Hamden, Conn.).

The best current research on the status and hope for recovery of the American chestnut in New England comes from Connecticut Agricultural Experiment Station publications and web pages.

Barron, Hal S. *Those Who Stayed Behind: Rural Society in Nineteenth-Century New England.* Cambridge, England: Cambridge University Press, 1984.

Describes the fate and history of New Englanders who chose to stay on their upland farms as most of the population moved away, to the expanding industrial and urban centers and to the midwestern United States.

Beattie, Mollie, Charles Thompson, and Lynn Levine. *Working with Your Woodland: A Landowner's Guide.* Hanover, N.H.: University Press of New England, 1993.

A straightforward and practical introduction to woodland management geared to the small woodlot owner.

Bidwell, Peter W., and James J. Falconer. *History of Agriculture in the Northern United States, 1620–1860.* Carnegie Institute Publication, no. 358. New York: Peter Smith, 1941.

One of the classic examinations of northeastern agriculture through two and a half centuries of change.

Carroll, Charles F. *The Timber Economy of Puritan New England.* Providence, R.I.: Brown University Press, 1974.

A very thorough examination of the importance of wood and wood products in the early period of New England growth.

Conuel, Thomas. *Quabbin, the Accidental Wilderness.* Rev. ed. Amherst, Mass.: University of Massachusetts Press, 1990.

The creation in the 1930s of the Quabbin Reservoir, the source of metropolitan Boston's water, inadvertently resulted in the protection of the single largest public area of water and land in southern New England. Using historical and modern photographs and sources, the author provides a delightful introduction to one of Massachusetts' most important natural areas.

Cronon, William. *Changes in the Land: Indians, Colonists and the Ecology of New England*. New York: Hill and Wang, 1983.

The contrast between Indian and colonial attitudes toward the land and the environment provides a contrast that this delightfully well-written volume uses to describe the historical changes in the New England landscape.

Day, Gordon M. "The Indian as an Ecological Factor in the Northeastern Forest." *Ecology* 34 (1953):329–346.

The classic and most-cited source of the notion that Indian burning was a major factor controlling forest patterns at the time of European settlement.

DeGraaf, Richard, and R. Miller. *Conservation of Faunal Diversity in Forested Habitats*. London: Chapman Hall, 1996.

A nice overview of wildlife populations and habitat in dynamic forested landscapes.

DeGraaf, Richard M., and D. D. Rudis. *New England Wildlife: Habitat, Natural History and Distribution*. USDA Forest Service General Technical Report NE-108. Broomall, Pa.: Northeastern Forest Experiment Station, 1986.

A comprehensive review of the habitat requirements and distribution of New England wildlife species.

DeGraaf, Richard M., et al. *New England Wildlife: Management of Forested Habitats*. USDA Forest Service General Technical Report NE-144. Broomall, Pa.: Northeastern Forest Experiment Station, 1992.

One of a series of well-illustrated publications, this volume provides an overview of the wildlife of New England, their habitat preferences and responses to different types of silvicultural management, and their historical responses to the physical and social changes in the New England landscape.

Dunwiddie, Peter. *Changing Landscapes: A Pictorial Field Guide to a Century of Change on Nantucket*. New Bedford, Mass.: Nantucket Conservation Foundation, 1992.

Using paired historical and modern photographs of the same locations, supplemented with the author's considerable historical and ecological research and insights, this book provides striking evidence for the historical transformation of the New England landscape and some of its consequences for conservation.

Dunwiddie, Peter, et al. "Old-Growth Forests of Southern New England, New York, and Pennsylvania." In M. B. Davis, ed., *Eastern Old-Growth Forests*. Washington, D.C.: Island Press, 1996, pp. 126–143.

A review of the status of old-growth forests with a summary of some of the notable ecological lessons that have emerged from research on these forests.

Dwight, Timothy. *Travels in New England and New York*. New Haven, 1821–22; reprint, Cambridge, Mass.: Harvard University Press, Belknap Press, 1969.

Written by an early president of Yale College, these essays provide a detailed and information-rich snapshot view of much of the New England landscape as it was undergoing its remarkable transformation to agricultural productivity.

Fisher, Richard T. "Second-Growth White Pine as Related to the Former Uses of the Land." *Journal of Forestry* 16 (1918):253–254.

One of Fisher's early works, which combines his understanding of forest and landscape history with his practical understanding of a valuable timber species.

———. "New England's Forests: Biological Factors." *American Geographical Society*, Special Publication no. 16 (1933):213–223.

An introduction to some of the writings of the first director of the Harvard Forest concerning the role of human history and natural processes that control forest patterns and dynamics in New England.

Foster, Charles H. W., ed. *Stepping Back to Look Forward: A History of the Massachusetts Forest*. Petersham, Mass.: Harvard Forest, 1998.

Written to celebrate the centennial of the forests and parks system in Massachusetts, this multiauthor volume provides an overview of the ecological, economic, social, and educational history of forests in the Commonwealth.

Foster, David R. *Thoreau's Country: Journey Through a Transformed Landscape*. Cambridge, Mass.: Harvard University Press, 1999.

Insights into the conservation and ecology of the New England landscape based on an interpretation of its history, using as a source the journal writings of Henry David Thoreau.

Foster, David R., and Emery R. Boose. "Hurricane Disturbance Regimes in Temperate and Tropical Forest Ecosystems." In M. Coutts, ed., *Wind Effects on Trees, Forests and Landscapes*. Cambridge, England: Cambridge University Press, 1994, pp. 305–339.

In many temperate- and tropical-forest ecosystems, wind damage from hurricanes, typhoons, or other types of storms represents one of the major natural disturbance processes. This article provides an overview of the wind disturbance regimes for parts of New England and the Caribbean, along with a general review of hurricane meteorology and ecological effects.

Foster, David. R., and Glenn Motzkin. "Ecology and Conservation in the Cultural Landscape of New England: Lessons from Nature's History." *Northeastern Naturalist* 5 (1999):111–126.

This article argues that the legacy of human activity is pervasive in most New England ecosystems and consequently that an awareness of history and the cultural aspect of landscapes is essential background for effective conservation planning and management.

Garrison, J. R. *Landscape and Material Life in Franklin County, Massachusetts, 1770–1860*. Knoxville: University of Tennessee Press, 1991.

A detailed examination of the economy and land-use activities of western Massachusetts during the expansion and height of agricultural activities.

Golodetz, Alisa D., and David R. Foster. "Land Protection in Central New England: Historical Development and Ecological Consequences." *Conservation Biology* 11 (1996):227–235.

One of a series of articles, based on research done at the Harvard Forest, that addresses the dynamics and conservation of forests with an emphasis on southern and central New England.

Irland, Lloyd. *The Northeast's Changing Forest*. Petersham, Mass.: Harvard Forest, 1999.

A wealth of information on the current status and history of forest areas and management issues across the northeastern United States.

Jorgensen, Neil. *A Guide to New England's Landscape*. Chester, Conn.: Globe Pequot Press, 1977.

A very informative and readable review of New England's landscape from a largely geological perspective.

———. *A Sierra Club Naturalist's Guide to Southern New England*. San Francisco: Sierra Club Books, 1978.

An introduction to the landscape and natural history of southern New England, written for the nonexpert in an accessible style.

Leahy, Christopher, John H. Mitchell, and Thomas Conuel. *The Nature of Massachusetts*. Reading, Mass.: Addison-Wesley, 1996.

Written to celebrate the centennial of the Massachusetts Audubon Society, this volume merges interesting natural history, beautiful watercolor illustrations, and perceptive insights on conservation issues into a nice overview of the state of habitats, species, communities, and land protection in Massachusetts.

MacCleery, Doug. *American Forests: A History of Resiliency and Recovery*. Washington, D.C.: U.S. Department of Agriculture, Forest Service, 1992.

The history of forest exploitation, agriculture, and regrowth, with an emphasis on the eastern United States, is told effectively through photographs, well-chosen illustrations and graphs, and accessible text.

Matthiessen, Peter. *Wildlife in America*. New York: Penguin, 1995.

A readable discussion of the changing attitudes of Americans toward wildlife and the resulting dynamics in major animal species.

McKibben, Bill. "An Explosion of Green." *The Atlantic Monthly*, April 1995, pp. 61–83.

A somewhat hyperbolic, though informative, celebration of the regrowth of the eastern forest and its significance as a major event in conservation history.

Merchant, Carolyn. *Ecological Revolutions: Nature, Gender, and Science in New England*. Chapel Hill: University of North Carolina Press, 1989.

One of America's most perceptive environmental historians provides an intriguing discussion of cultural and ecological changes in the history of New England and identifies major "revolutions" in social and economic conditions that contribute to substantial changes in the state of the environment.

O'Keefe, John, and David R. Foster. "Ecological History of Massachusetts Forests." In Charles H. W. Foster, ed., *Stepping Back to Look Forward: A History of the Massachusetts Forests*. Petersham, Mass.: Harvard Forest, 1998, pp. 19–66.

A ten-thousand-year consideration of the dynamics of the New England landscape and the factors that control it.

Orwig, David A., and David R. Foster. "Forest Response to Introduced Hemlock Woolly Adelgid in Southern New England, USA." *Bulletin of the Torrey Botanical Club* 125 (1998):59–72.

The hemlock woolly adelgid is one of the most recent and potentially most destructive introduced pathogens to affect the New England landscape. This article reviews the initial impacts and consequences of hemlock mortality in natural forests across southern New England.

Patterson, William A., and Kenneth E. Sassaman. "Indian Fires in the Prehistory of New England." In G. P. Nichols, ed., *Holocene Human Ecology in Northeastern North America*. New York: Plenum Publishing Company, 1988, pp. 107–135.

Using paleoecological, historical, and archaeological studies, this article considers the role that Indian populations may have had in controlling vegetation patterns through the use of fire and related subsistence activities.

Peterken, George F. *Natural Woodland: Ecology and Conservation in Northern Temperate Regions*. Cambridge, England: Cambridge University Press, 1996.

An unusual and highly informative text on forest conservation based on consideration of woodland characteristics in North America and Europe, including nice descriptions of natural vegetation and forest processes and a wide array of forest-management practices.

Pinchot, Gifford, and Henry S. Graves. *The White Pine*. New York: Century, 1896.

A classic early paper on one of eastern North America's most important tree and timber species by two of America's earliest foresters.

Pyne, Steve. *Fire in America: A Cultural History of Wildland and Rural Fire*. Princeton, N.J.: Princeton University Press, 1982.

A wide-ranging and informative consideration of the natural and anthropogenic role of fire in North America.

Raup, Hugh M. "The View from John Sanderson's Farm." *Forest History* 10 (1966):2–11.

Based on a popular lecture by Raup that focused on the history of the Prospect Hill tract of the Harvard Forest, this article effectively presents a human picture of the changing land-use practices that have shaped much of the New England countryside over the past 300 years.

Robinson, William F. *Abandoned New England: Its Hidden Ruins and Where to Find Them*. Boston: New York Graphic Society, 1976; Robinson, W. F. *Mountain New England: Life Past and Present*. Boston: Little, Brown and Co., 1988.

Two beautiful but unfortunately hard-to-find books that provide a wonderful introduction to the history of the New England landscape and approaches to finding traces of this history in the modern landscape.

Russell, Emily W. B. *People and the Land Through Time: Linking Ecology and History*. New Haven: Yale University Press, 1997.

A historical ecologist uses her own travels in Great Britain and research from the eastern United States to underscore the importance of considering the role and effect of human activity on modern landscapes and to review methodological approaches for pursuing landscape history.

Russell, Harold S. *A Long, Deep Furrow: Three Centuries of Farming in New England*. Hanover, N.H., and London: University Press of New England, 1976.

A wide-ranging and very well researched overview of the changes in New England's agriculture.

Smith, David M. *The Practice of Silviculture*. 9th ed. New York: John Wiley and Sons, 1996.

The completely revised edition of the classic text on temperate-forest silviculture.

Spurr, Steve H., and Albert C. Cline. "Ecological Forestry in Central New England." *Journal of Forestry* 40 (1942):418–420.

Based on the silvicultural and land-use studies of Richard T. Fisher and other Harvard Forest researchers, this paper provides a remarkable new perspective on forest management in the first half of the twentieth century.

Thomson, Betty F. *The Changing Face of New England*. New York: Macmillan, 1958.

One of New England's foremost plant anatomists took a break from her own studies and teaching to provide a delightful introduction to the natural history and changing character of the landscape of the northeastern United States.

Vickery, Peter, and Peter Dunwiddie. *Native Grassland of the Northeastern United States*. Lincoln, Mass.: Massachusetts Audubon Society, Center for Biological Conservation, 1998.

In the eastern United States, are grasslands a naturally occurring habitat or are they artificial constructs of human history? This book reviews current thinking on the history and conservation of these threatened parts of the landscape, as well as the status of other current habitats, species, and communities.

Wessels, Tom. *Reading the Forested Landscape: A Natural History of New England*. Woodstock, Vt.: Countryman Press, 1997.

This book provides people who are interested in interpreting forest history on their woodland walks with a nice introduction to some of the cultural and biological clues frequently encountered in the New England countryside, supplemented by dark stylistic illustrations.

Whitney, Gordon. G. *From Coastal Wilderness to Fruited Plain: A History of Environmental Change in Temperate North America from 1500 to the Present*. Cambridge, England: Cambridge University Press, 1994.

The definitive environmental history of the northeastern United States is supported by an exhaustive bibliography and new and integrative figures and tables. This volume describes the early forest vegetation and then documents its changes through the early twentieth century and provides an authoritative introduction to the scholarly methodology that is the toolbox for historical ecology.

Williams, Michael. *Americans and Their Forests: A Historical Geography*. Cambridge: Cambridge University Press, 1989.

A wide-ranging consideration of the forest history of the United States, with very informative data on the forest industry and forest statistics.

ABOUT THE HARVARD FOREST

Since its establishment in 1907 the Harvard Forest has served as a center for research and education in forest biology. Through the years researchers at the Forest have focused on silviculture and forest management, soils and the development of forest site concepts, the biology of temperate and tropical trees, forest ecology, forest economics, and ecosystem dynamics. Today, this legacy of research and education continues as faculty, staff, and students continue to interpret historical and modern changes in the forests of New England and beyond resulting from human and natural disturbance processes, and to apply this information to the conservation, management, and appreciation of forest ecosystems.

Physically, the Harvard Forest comprises approximately 3,000 acres of land in Petersham, Massachusetts that include mixed hardwood and conifer forests, ponds, extensive spruce and maple swamps, and diverse plantations. Additional land holdings include the 25-acre Pisgah Tract in southwestern New Hampshire (located in the 13,000-acre Pisgah State Park), a virgin forest of white pine and hemlock, which was 300 years old when it blew down in the 1938 Hurricane; the 100-acre Matthews Plantation in Hamilton, Massachusetts, which is largely made up of plantations and upland forest; and the 90-acre Tall Timbers Forest in Royalston, Massachusetts. In Petersham a complex of buildings that includes Shaler Hall, the Torrey Laboratories, and the Fisher Museum provide office and laboratory space, computer and greenhouse facilities, and a lecture room and lodging for seminars and conferences. Nine houses provide accommodation for staff, visiting researchers, and students. Extensive records including long-term data sets, historical information, original field notes, maps, photographic collections and electronic data are maintained in the Harvard Forest Archives.

The Harvard Forest is part of the Faculty of Arts and Sciences of Harvard University and administers the Graduate Program in Forestry that awards a master's degree in Forest Science. During the summer months the Harvard Forest Research Program offers opportunities for undergraduates from Harvard and other colleges nationwide to pursue ecological studies independently or as part of major research teams.

THE FISHER MUSEUM

The Fisher Museum, dedicated in 1941, was constructed to exhibit the dioramas and provide a means of conveying the results of research at Harvard Forest to the scientific community and the general public. Renovations in 1972 provided additional exhibit space on the second floor and created an auditorium surrounded by the dioramas, which became the Ernest M. Gould Audiovisual Education Center in 1989. Using the dioramas as a focus, additional Museum exhibits expand the interpretation of land-use history, forest ecology and development, and the effects of major disturbances such as the 1938 hurricane, and place this interpretation within the context of ongoing research at Harvard Forest.

Throughout the year the Fisher Museum provides informal education to a wide range of students at all levels, from elementary through graduate, as well as to professional groups and the general public. Self-guided nature trails through the adjacent forest extend the lessons from the dioramas and Museum exhibits into the ever-changing current landscape and allow all visitors an opportunity to learn more about the history and ecology of New England forests and how they are being studied.